Reg's

Practical

Guide

To

Using Your

Android

Phone

By Reginald T. Prior

Visit http://www.rcsbooks.com or E-Mail me at reginaldprior@rcsbooks.com

Printed in the United States of America

First Printing: September 2010

ISBN - 1453851321
EAN - 139781453851326

Trademarks And Copyrights

Trademarked and or copyrighted names appear throughout this book. Rather than list and name the entities, names or companies that own the trademark and or copyright or insert a trademark or copyright symbol for with every mention of the trademarked and or copyrighted name, The publisher and the author states that it is using the names for editorial purposes only and to benefit the trademark and or copyright owner, with no intentions of infringing on the trademark and or copyrights.

Warning and Disclaimer

Every effort has been made to make this book as complete and accurate as possible. No warranties are implied. The information provided is on a "as is" basis. The author and the publisher have no liability or responsibility to any individuals or entities with any respect to any loss or damages from the information provided in this book.

Preface

There are many books on the market that teach people how to use technology. But as I look through many of these books, I have found that they teach some of basics, but miss a lot of critical things about how to fully utilize technology.

My aim of this book is to fill in these gaps that most books don't cover or spend sufficient time covering in common sense and in a way that is easily understood by everyone. As a computer technician for 12 years, I've come across many people that understand some things about technology, but want to have a better understanding about how it works and how to fully utilize them in their everyday lives.

In this book, I will be covering Google's Android Operating System. At the time of the writing this book, this is the latest operating system that many cellular phone carriers such as Verizon, Sprint, T-Mobile and many others are currently producing cell phones using this operating system.

Google's Android platform is very powerful, simple and flexible enough to compete and in some cases outdo Apple's IPhone in some aspects. We will go into that and more in this book.

You as the reader are the most important critics of this book. I value all of your feedback and suggestions that you may have for future books and other things that I can do to make these books better. You can e-mail me at reginaldprior@rcsbooks.com and please include the book title, as well as your name and e-mail address. I will review your comments and suggestions and will keep these things in mind when I write future texts. Thank you in advance,

Reginald T. Prior

Acknowledgements

This book that you are reading right now takes a lot of time and sacrifice to put together. I would first and foremost thank God for giving me at the age of six the love of working on technology that still is as strong today as it was back then. I would like to thank my wife, Sharifa for being a trooper when I was spending many hours on my laptop putting this book together and also for being there to help me read my drafts to make sure that it would be understood.

Also I would like to thank my family and many friends that helped and supported me throughout the years on many other projects and being there for me in good times and bad. I hope that you all enjoy this book as much as I had putting it together.

Hello Everyone,

I would like to thank you in advance for purchasing "Reg's Practical Guide To Using Your Android Phone" I aim to make the Google Android operating system easy to learn while showing you the many features that Google has implemented to make Android the choice of many cell phone carriers for Smartphones.

Table Of Contents:

Chapter One: Touch Screen Phone Terminology Dictionary

Touch Screen Phone Terminology Dictionary ----------- 11

How To Turn Your Android Phone On And Off -------- 19

Chapter Two: Getting To Know Your Android Phone

What Does All Of These Buttons On This Phone Do? --23

How To Tell What Version Of Android You Have -------26

Navigating The Main Android Screen ----------------------28

Making A Call --32

Answering or Denying A Call ---------------------------------35

Setting Options Through The Settings Menu ------------36

Putting Phone Into Silent Mode ------------------------------39

Changing The Ringtone On Your Phone -------------------40

Setting Up Bluetooth Devices ---------------------------------56

Adding People And Phone Numbers To Your Phone ---58

Deleting Contacts Off Your Phone --------------------------67

How To Send A Text To Someone --------------------------68

How To Get Onto The Internet ---------------------------------73

How To Setup And Use E-Mail --------------------------------80

Sending Email From Your Phone --------------------------98

Deleting Email From Your Phone --------------------------101

Using The Camera To Take Pictures And Video --------103

How To Download & View Pictures On Your Phone --107

How To Download Music To Your Phone ---------------119

Using The Gallery To Delete Pictures & Video----------130

Using The Music App To View And Play Your Music --137

Chapter Three: Using The Android Market To Add Functionality To Your Android Phone

Getting To & Looking Through The Android Market -143

Downloading And Installing Apps ----------------------- 144

Uninstalling Apps That You Don't Use -------------------151

Chapter Four: Smartphone Security

The Three Things You Can Do To Protect Your Data And Phone From Cyber Thieves And Other Threats--------156

Chapter One:

Touch Screen Phone Terminology Dictionary

Chapter One: Smartphone Terminology Dictionary

Before we go into learning how to operate Android or any other Smartphone, I believe that before you can have knowledge about anything, you have to build your knowledge like a builder builds a house. First you have to lay a solid foundation down before we can start building floors and rooms within the house. Understanding Smartphone terminology and what it means is like laying the foundation on the house.

This chapter translates what we geeks talk about when we are talking about Smartphones. Decode the foreign language so to speak. This is not a full list of Smartphone terminology, but this is a list of the most common terminology used for talking about most Smartphone's. So with no further delays, let's get started laying the foundation to being confident to use your Android Smartphone.

Note - When you go to purchase an Android phone, you would have to have a Gmail account (where you can go to http://www.gmail.com to get an FREE account), or your cell phone provider will sign you up for one when you are purchasing an Android phone.

Smartphone –

Smart phones are a category of mobile device that provides advanced capabilities beyond a typical mobile phone. Smartphone's run complete operating system software that provides a standardized look and feel.

SMS –

SMS stands for *short message service*. SMS is also often referred to as texting, sending text messages or text messaging. The service allows for short text messages to be sent from one cell phone to another cell phone or from the Web to another cell phone.

MMS –

MMS Messaging stands for *multimedia messaging service*, takes SMS (*short message service*) text messaging a step further by allowing pictures and video to be attached to text messages and allows longer lengths of texts beyond the traditional, 160-character SMS limit.

QWERTY –

The term that commonly describes today's standard keyboard layout on English-language computers.

Airplane Mode –

AKA offline, radios off, or standalone mode. Some phones and other wireless devices have a special "flight" or "airplane" mode that turns off just the wireless radio parts of the device, for safe use on an airplane where radio transmitters are not allowed.

Accelerometer-

A component that measures tilt and motion. A device with an accelerometer knows what angle it is being held at. It can also measure movements such as rotation, and motion gestures such as swinging, shaking, and flicking.

One common use in phones it to detect whether the phone is upright or sideways and automatically rotate the graphics on the screen accordingly. Another common use is controlling games and other applications (such as music player) by moving or shaking the phone.

Left And Right Swipe –

To touch the screen of the phone and sliding your finger to the left for a Left Swipe, or to the right for a Right Swipe. Commonly used to navigate between screens on your phone.

The next pictures show and example of both left and right swiping.

Left Swiping

Right Swiping

Up And Down Swipe –

To touch the screen of the phone and sliding your finger up for an Up Swipe, or down for a Down Swipe. Commonly used to navigate, show more options on a menu, or to navigate more of the webpage on your browser.

Single/Double Tap –

To touch the screen of the phone and quickly release once for single tap, twice for double tap. Commonly used to select or activate something. (Think of this as single or double clicking the mouse button on your computer)

D-Pad –

A D-pad (Directional Pad) is a set of four buttons arranged in a circle or diamond that function as "arrow" keys, allowing the user to move up, down, left, and right within the device's user interface, such as scrolling through menus. Many D-pads also have a button in the center that performs a "select" or "OK" function.

Geo-Tagging-

Geo-tagging is associating a geographic location with an item such as a photo. Some phones with both a camera and GPS can record the precise location a photo was taken and automatically embed that location information into the photo file.

GPS –

GPS (Global Positioning System) is a global satellite-based system for determining precise location on Earth. In a phone, this will allow operators to immediately receive your location when you call the emergency number (911 or 112).

Multi-Touch –

The ability for a touch surface to respond to multiple finger touches at the same time. For example, a common use for multi-touch is pinch-to-zoom, where you can place two fingers on a screen image (like photo, web page, or map) and spread your fingers to zoom in, or pinch your fingers together to zoom out.

Ringtone –

A sound that a phone makes for an incoming call. All modern mobile phones allow at least several choices for different sounds or melodies for a ringtone. Most phones also allow additional new ringtones to be downloaded or sent to phone, for added personalization.

Voice Mail –

A service provided by your cell phone company to store and manages voice messages for individual users. Like an answering machine, voice mail can handle a call when the person being called is unavailable, by playing a greeting message and recording a voice message from the caller.

3G-

3G Stands for 3rd-generation. Analog cellular phones were the first generation. Digital phones marked the second generation (2G). 3G is loosely defined, but generally includes high data speeds, streaming video and always-on data access, and greater voice capacity.

4G –

A somewhat vague term used to describe wireless mobile radio technologies that offer faster data rates than current 3G (third generation) technologies.

Turning On And Off Your Android Phone

The first thing that has to be done before we can start working with an Android phone is to have your phone turned on. Every Android phone is made differently and phone manufacturers place the power button in many different places. But the one thing you need to keep in mind is that the power button has an icon that looks like the picture below:

To turn your phone on, find the button that has the icon that looks like the picture shown above and press and hold down that button. After about 5 seconds, your phone should turn on and then you can let go of that button.

To turn your Android Phone off, you will press and hold down the same button, and a menu choice will come up looking like the next picture:

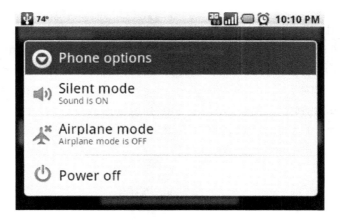

Single tap the "Power Off" option. Another popup will come up saying that the phone will shut down. Single tap the "OK" button and then the phone will turn off.

Chapter Two:

Getting To Know Your Android Phone

Over the years, trends in technology have changed. One trend was the shift from using desktop computers to notebook computers. This change happened because people had started to be more mobile and needed to use a computer on the go. So notebook computers became more popular than a desktop.

But for the past year or so, technology is currently in process of another trend happen. Cellular phones (smartphones in particular) are suddenly able to do as much or more than a notebook computer, but in a smaller package. So more people are now swapping their notebooks for these powerful cellular phones.

This trend is going to continue to explode due to the fact that getting a smartphone is cheaper than buying a notebook computer and can do all of the same things as a notebook like sending and receiving email, playing games, surfing the Internet, installing applications as well as texting and making phone calls.

After you have pushed the power button on your Android phone and turned it on, your phone will do a quick hardware check. After that check has completed, the phone starts a "Master Program" called the operating system. That process is called booting up.

The booting up process is the act of the phone setting up everything it needs to operate. Once Android has fully booted up, it will go right to the main screen where you swipe right on the lock icon to unlock to start using your phone. The picture below shows what that screen looks like.

Before we go into actually using your Android phone, we have to go over the buttons on your phone and what they are used for, because you will use them a lot. These are pictures are from a LG Ally phone, but all Android phones have similar buttons.

-(On the left side of your phone) This button is for answering phone calls on your Android phone when you receive a phone call. This is primarily the only function of this button on your Android phone.

- This is the back button on your Android phone. This button is used for multiple functions. One function of this button is to go back to a previous screen within the phone or a previous webpage when surfing the Internet.

- This is the home button on your Android phone. This button is used to take you back to the main screen from anywhere.

- This is the menu key. This is one of the most important functionality keys on the Android phone. This key is most commonly used to provide additional options for your phone or within an application. Think of this button as your "File" menu within any Windows program.

- (On the right side of your phone) This button is for ending phone calls on your Android phone. Also on some phones, this button is used to turn on or off your phone simply by pressing and holding it down. Check your phones user manual for the location of the power button.

24

 - This button is your web search button. You press this button and type in the dialog box to do a web search on Google.

Quick Tip - If you press and hold this key down, A window will pop up telling you to "Speak now." Do so, and Google will automagically search for whatever you said. The next picture shows what this screen looks like.

At the time of the writing of this book, there are two versions of Android operating system that are installed on cell phones. There is version 2.1 and version 2.2. Both versions are covered in this book. Most functions work in 2.1 and in 2.2, but explanations are made when there are differences between the two versions.

You might be asking yourself, how do I know what version of Android my particular phone has? On the main screen (You would get to the main screen by pressing the "Home" key on your phone. It would look like this), you would press the menu key, which looks like this and the options will come up like the next picture:

Single tap the "Settings" option and the settings menu will come up like the next pictures:

Swipe up and Single tap the "About Phone" option and the screen should look like the next picture:

Look at the Android Version line in this menu. You will either see a 2.2 or 2.1 in this line. Whatever your phone displays in this line is the version of Android your phone presently is running. After you have checked the version of Android, Press the home key to go back to the main screen.

Navigating The Main Android Screen

After you have right swiped to unlock your Android phone, you will be presented with the main screen as shown in the next picture.

In the next few sections, we will go over in detail about everything you will see on the main screen of an Android phone. The next picture shows the most important things to keep in mind when you look at the main screen on your Android phone:

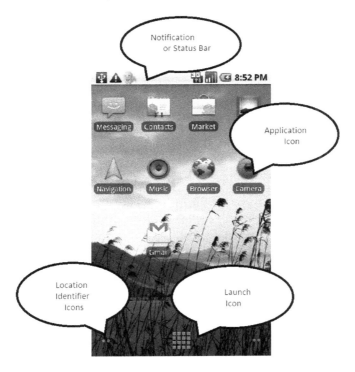

- The Notification or Status Bar shows information including the time, battery status, signal strength and other information. The next picture shows all of the Notification Icons that also will show up here also.

[M]	New Gmail message	[icon]	3 more notifications not displayed	
[icon]	New text or multimedia message	[icon]	Call in progress	
[!]	Problem with text or multimedia message delivery	[icon]	Call in progress using Bluetooth head-set	
talk	New Google Talk™ message	[icon]	Missed call	
[oo]	New voicemail	[icon]	Call on hold	
[1]	Upcoming event	[icon]	Call forwarding is on	
[icon]	Data is syncing	[icon]	Uploading data	
[!]	Problem with sign-in or sync	[icon]	Downloading data	
[icon]	microSD card is full	[icon]	Download finished	
[icon]	An open Wi-Fi network is available	[icon]	Connected to VPN	
[icon]	Phone is connected via USB cable	[icon]	Disconnected from VPN	
[►]	Song is playing			

- The Application Icon allows you single tap an icon to start a specific Application. Think about this as a desktop shortcut on your Windows or Macintosh computer.

Note – You are not limited to using the icons that your phone comes with by default. You can add or remove icons from any one of your home screens. The following sections explain this in detail.

To Remove an Application Icon From Your Home Screen:

1. Touch and hold the application icon you wish to remove from the home screen. The phone will vibrate and you should notice a trash can located at the bottom of the screen.

2. Drag the icon with your finger to the trash can and let go of the icon when the icon turn a red color. The icon is now removed from your home screen

To Add an Application Icon To Your Home Screen:

1. Single tap the launch icon on the home screen (You would get to home screen by pressing the home key on your phone). Swipe up or down to find the application you want to put on your home screen.

2. Touch and hold the desired application you want on your home screen. The phone will vibrate and your home screen will appear. Let go the icon to add the application to your home screen.

- The launch icon allows you to single tap to on it to show all of the applications that are currently installed on your phone.

- The Location identifier dots allow you a quick glance to see which home screen you are at within Android. I will explain more in the note section.

Note – You can have multiple home screens on an Android phone. Most phones allow you to have at least 5 home screens where you can add application icons, or Widgets to show different kinds of information.

To go to another home screen, just left or right swipe on the home screen and your phone will transition to the next home screen. To get back to the main home screen, just press the home key on your phone that looks like this:

Making A Call On Your Phone

The one main thing that you want to do on your Android Phone is to make a phone call right? Since this is a smartphone, buttons and other functions are a little bit different than on a regular cellular phone.

To make a phone call on Android, Single tap the launch icon and swipe up or down to find the phone application. The icon looks like the next picture

Note – In Android Version 2.2, the phone application is called the dialer with a similar looking icon.

Single tap on the icon and the phone application will then come up and look like the next picture:

Use the touchscreen to dial the number you wish to call and single tap the green call button which looks like the next picture:

Then your phone will dial the number you have just entered into the phone and the screen will look like the next picture:

To end the call, single tap the red "End" button or press the end call button which looks like this , and the phone will end the current call.

Answering/Denying A Call On Your Phone

You are doing something and your phone suddenly rings. And you look at your phone and find that answering a phone call on your Android phone is a little different from a regular cellular phone. When a phone call comes in, the screen of your Android phone will look similar to the next picture

To answer the call, swipe right the answer icon. To send the caller to voicemail, swipe left the red deny icon.

Setting Phone Options Through The Settings Menu

There are situations where you have to put your phone on vibrate mode, change the ringtone or do anything to change the way your phone operates. You would go to the settings menu to accomplish this. The settings menu at first glance can be very intimidating because there are so many options. But we only need to be concerned with a couple of selections at this time.

To get to the settings menu, on the main screen (You would get to the main screen by pressing the "Home" key on your phone. It would look like this), you would press the menu key, which looks like this and the options will come up like the next picture:

Single tap the "Settings" option and the settings menu will come up like the next pictures:

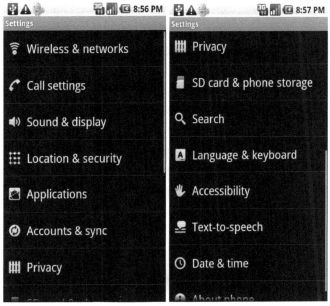

Settings (Part One)   Settings (Part Two)

Like I had mentioned earlier in this section, the settings menu can be intimidating at first glance, but we need to only at this time be concerned with only a few options here:

1. Sound & Display (In Android 2.2, It is the sound option) – This is the option where you can adjust the ringer volume, media volume, change your ringtone or put your phone into silent mode.

2. Wireless & Network Settings – This is the section where you would put your phone into "Airplane Mode", set up options for what the phone would do when roaming, configure Bluetooth devices such as hands free earpieces or connect to local Wi-Fi hotspots.

We are about to go into more detail of the options in the Sound & Display menu option. Single tap that menu selection to go into that menu. Your screen should look like the next picture:

In this menu, a lot of the menu choices are self-explanatory, Such as putting the phone into silent mode. The way that you would do that is by single tapping that option and a green check will be placed in the box. Then the phone is in silent mode. To take it out of silent mode, just single tap the silent mode option again, and the green check will go away. Then the phone is back in normal operating mode.

To adjust the volume of the ringer, single tap the Ringer Volume option and the ringer volume slider menu will look like the next picture. (In Android 2.2, it is called the Volume option), and also in this menu, you can adjust the ringtone and media volume as shown in the picture.

 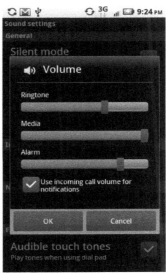

Android 2.1 Menu   Android 2.2 Menu

Left or right swipe the slider to adjust the ringer volume and single tap the "OK" button to set the volume levels. You would do the same exercise for setting the media volume, which controls the playback volume for music and videos.

To change the ringtone for your phone, single tap the "Phone Ringtone" option and a dialog box will come up, showing all of the ringtones that are available on your phone. It will look like the next picture:

Swipe up or down to look at the selection of ringtones. Select a ringtone by single tapping on it and the sample sound will play. If you like it, single tap the "OK" button and the ringtone will be set on your phone. If not, single tap the "Cancel" button.

Setting A Separate Ringtone For A Person In Your Contact List

You may have just set up a ringtone for all of your incoming calls, but you may want a specific person on your contact list to have a separate ringtone for when they call you. You can do that on your Android phone. Here's how you would do it.

Single tap the launch icon from the main screen (You would get to the main screen by pressing the home key on your phone. It look like this) swipe up or down to find the Contacts application. Single tap the Contacts app and the screen will come up looking like the next picture

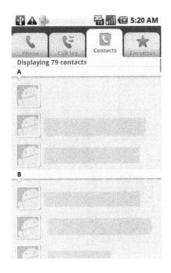

Swipe up or down to find the person that you want to set a separate ringtone for and single tap on that person to bring up their information. Press the menu key, which looks like this  and the options will come up like the next picture:

Single tap the options choice and your screen should look like the next picture:

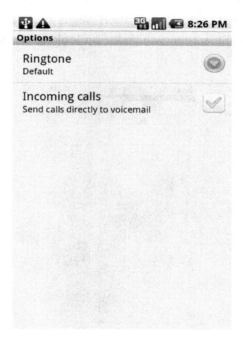

Single tap the Ringtone menu choice and the list of available ringtones that are on your phone will come up like the next picture:

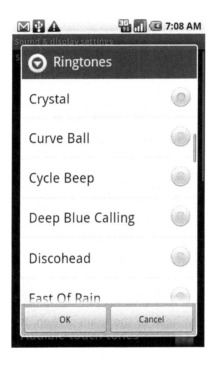

Swipe up or down through the ringtones. Single tap a ringtone and the phone will play the sample sound. If you like this ringtone for this contact, then single tap the "OK" button and the ringtone will be set for this contact as shown in the next picture:

The ringtone "BentleyDubs" will be set for this person.

Press the home key that looks like this 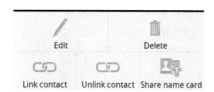, and you will be brought back to the main screen. In Android 2.2, to change the ringtone for a specific contact, when you have the person's information open, press the menu key , and the menu will come up like the next picture:

Single tap the "Edit" menu choice. Your screen should look like the next picture:

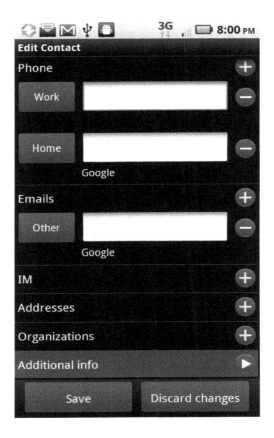

Single tap the "Additional Info" menu choice to open the menu. Swipe up all the way until you see an option for ringtone. Your screen should look like the next picture:

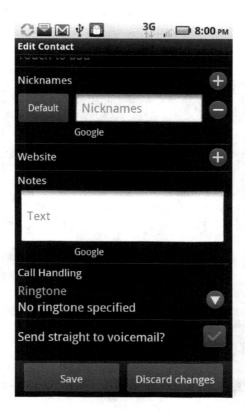

Single tap the "Ringtone" menu choice and a menu with all of the available ringtones on your phone will come up and look similar to the next picture:

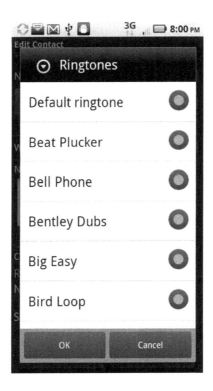

Swipe up or down through the ringtones. Single tap a ringtone and the phone will play the sample sound. If you like this ringtone for this contact, then single tap the "OK" button and the ringtone will be set for this contact as shown in the next picture:

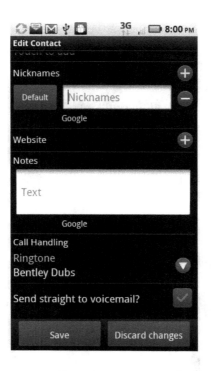

As you can see, The Ringtone "Bentley Dubs" is set for this contact. To save the ringtone for this contact, single tap the save button and the ringtone will be set for this contact. Go back to the main screen by pressing the home button 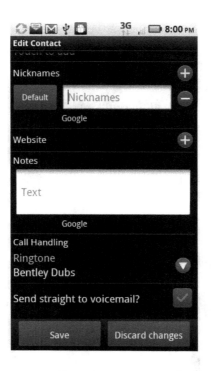 on your phone.

The next set of phone options that we will cover at this time is the "Wireless & Network Settings" section. You would get to this section by getting to the phone settings menu.

On the main screen (You would get to the main screen by pressing the "Home" key on your phone. It would look like this 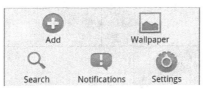), press the menu key, which looks like this ▬ and the options will come up like the next picture:

Single tap the "Settings" option and the settings menu will come up like the next picture:

Phone Settings Menu

Single tap the "Wireless & networks" option and the screen should look like the next picture:

In this menu, the choices and the phrases that are used here may be new to you, so I am going to explain what they mean and why you should know about them.

Airplane Mode – AKA offline, radios off, or standalone mode. Some phones and other wireless devices have a special "flight" or "airplane" mode that turns off just the wireless radio parts of the device, for safe use on an airplane where radio transmitters are not allowed.

To set your phone on Airplane mode, single tap the "Airplane Mode" menu choice, and a green check mark should show up like the next picture:

Showing Airplane mode is active. To turn Airplane Mode off, then single tap the menu option again and the check mark will go away.

Wi-Fi Mode – Your cellular phone is capable of connecting to Wi-Fi hot spots as well as the Internet connection provided by your cell phone carrier. The advantage of this is that you can enjoy the same high speed wireless lines that you would on your laptop.

Also when you are downloading updates to your phone (Which we would go over In the Android Market chapter) or music files from Amazon, downloads would be faster than it would be strictly on your wireless carrier's 3 or 4G wireless connection.

Also, if don't have a unlimited data plan (Wireless carriers has pretty much eliminated this practice), then you could enjoy using your phone for surfing the Internet without fear of overage charges.

To switch your phone onto WI-FI mode, single tap the "Wi-Fi Mode" menu choice, and a green check mark should show up like the next picture:

After you have turned on Wi-Fi mode, to view wireless access points in your general area, single tap the "Wi-Fi Settings" menu option. Your phone will scan for Wi-Fi access points that are nearby and list them in this menu. Your screen should look similar to the next picture.

To connect to a particular access point, single tap on that access point label, and if it requires a password, then the phone will prompt for a password as shown in the next picture:

Single tap the textbox and the touch keyboard will then come up. Type in the password and single tap the "Done" key on the touch keyboard to close it. Next single tap the "Connect" button. Your phone should then connect to the access point. When the phone has successfully connected, you should see this icon on the notification bar.

Also in this menu, if you have possession of a wireless Bluetooth earpiece that looks similar to the next picture:

55

You can connect (Or Pair) to your phone on this menu. And to do that, you will have to have your Bluetooth setting on your phone turned on. To do that, single tap the Bluetooth menu choice to put a green check mark in the box. When you do that, your screen should look like the next picture:

When the Bluetooth settings are turned on, read the instructions of your particular earpiece on how to set it to pairing mode. When your earpiece is in pairing mode, then single tap the "Bluetooth Settings" menu choice and your screen should look like the next picture:

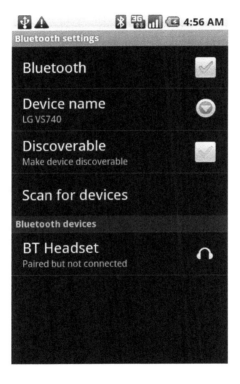

At this point, you should be able to single tap the "Scan For Devices" menu choice, and the phone will scan for your earpiece.

Note – For most earpieces, the pairing password should be the same and the phone should take care of the pairing. But if your password is different, then the phone will ask for the password to the earpiece. Then use the touch keyboard to enter the password and then you should be all set.

Adding People And Phone Numbers To Your Phone

To add people to your address book on your phone, Single tap the launch icon from the main screen (You would get to the main screen by pressing the home key on your phone. It look like this) and swipe up or down to find the Contacts application. Single tap the Contacts app and the screen will come up looking like the next picture:

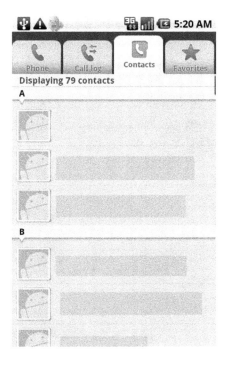

To enter a new contact into the phone, press the menu button on the phone which looks like this:  and the menu for the contacts application will come up and look like the picture shown below:

59

Single tap the "New Contact" option under this menu, and the new contact menu screen will come up like the picture shown below:

To start entering information for your new contact, single tap the "Given Name" field and then the touch keyboard will come up as shown in the next picture:

Type in the first name of your contact in this field, then swipe up slowly and single tap the "Family Name" text field to type in the last name of this contact. Depending on the screen size of your particular phone, you may have to swipe up to show the part of the menu where you will enter the phone number of this contact.

In Android 2.1, Single tap the field of the type of phone number that you want to enter. For example, if you were entering the cellular phone number of this contact, single tap the mobile text field. When you single tap the phone number field, the number pad will show up as shown in the next picture

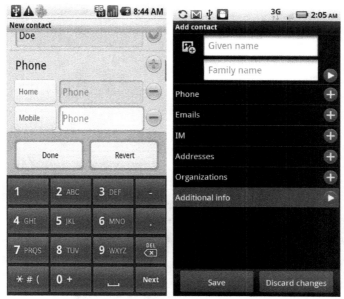

Android 2.1 Screen   Android 2.2 Screen

Use the keypad to type in the phone number of this contact into the text field.

In Android 2.2, you would have to single tap the plus symbol to add a phone number for this person. A new line will come up to enter the home phone number for this contact. When you single tap the phone number field, the number pad will show up so that you can enter the phone number of this contact.

A Couple of Notes –

- You are not limited to entering just the basic type of information for this user. You can add additional information for this contact like their E-Mail, Home and Business Addresses and other information as you can see in the picture shown below:

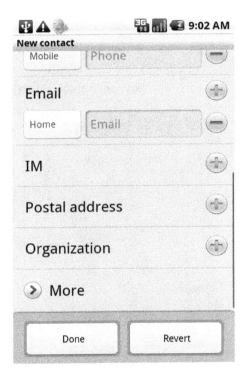

- To add more than one type of phone number, address or any other kind of information, locate and single tap the green plus symbol that looks like this , and another line will show up allowing you to add additional information.

- To enter a different type of information for this person for example, you have just entered their home phone number, but want to also add his work number. To do that, single tap the green plus symbol to add another text field to enter the number as shown in the picture below:

In this case, the new line comes up showing "Mobile", but you want to change that to "Work". To change this, single tap on the word "Mobile", then the label options menu will come up as shown in the picture below:

Single tap the "Work" label. The label for that text field will now be changed to "Work" as shown in the picture on the next page:

Then single tap the text field and enter this contacts work number. When finished entering all of the information for this contact, single tap the "Done" button. Your contact and all of their information you have entered will now be saved on your phone.

Deleting Contacts Off Your Phone

To delete a contact, Single tap the launch icon from the main screen (You would get to the main screen by pressing the home key on your phone. It look like this) swipe up or down to find the Contacts application that looks like the picture shown below:

Single tap the icon. When the contacts screen comes up, Swipe up or down to find the contact that you want to delete from the phone. Single tap the contact to pull up that person. Press the menu button on your phone. The options menu will come up as shown in the picture below:

To delete this contact, single tap the "Delete Contact" option. A dialog box will come up stating that this contact will be deleted from this phone. Single tap the "OK" button and the contact will be deleted from your phone.

67

How To Send A Text To Someone

To send a text to another person's cell phone, Single tap the launch icon from the main screen (You would get to the main screen by pressing the home key on your phone. It look like this) swipe up or down to find the messaging application. Find the messaging icon which looks like the next picture,

And single tap on it (In Android 2.2, the texting application is called "Text Messaging"). The messaging app will come up looking like the next picture:

Next single tap the "New Message" link to pull up the texting window as shown in the next picture:

Single tap the "To" box and the touch keyboard will come up on the screen where you would type in the cellular phone number of the person you want to send the text to (Area Code first).

Note – You would have to single tap on the "?123" key to display the numbers keypad where you can type numbers in the "To" textbox.

If you have some cellular contacts already in your address book, they should show up where you can swipe up or down and single tap a specific person as shown in the next picture:

Once you entered the phone number of the person you are sending this text message to, single tap the "Type To Compose" text box and type your message. When you are done typing the message, single tap the "Send" button. Your screen should look like the next picture:

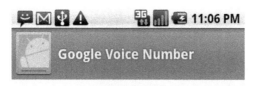

In this case for an example, I sent a text to my Google voice number that is also attached to my phone. As you can see, that is how you would text someone.

How To Get Onto The Internet

One of the best aspects of having a smartphone is having the ability to surf the internet just about the same way as you would on a computer or a laptop. On other types of smartphones, you would get the Mobile or condensed versions of websites because the browsers on those phones couldn't handle the layouts of those webpages.

But with Android, the browser app shows websites in their full glory just like if you were on your computer as shown in the next picture:

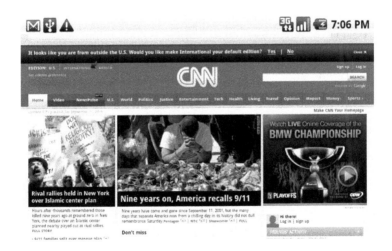

To go on the Internet, Single tap the launch icon from the main screen (You would get to the main screen by pressing the home key on your phone. It look like this) and swipe up or down to find the browser application. Single tap the browser app icon. The icon looks like the next picture:

The Android web browser will come up like the next picture:

To bring up the address bar so that you can go to a different webpage than the one that comes up by default, swipe down to go to the top of the current webpage, and the address bar will come up as shown in the next picture:

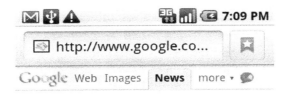

Single tap inside the address bar and the touch keyboard will come up where you can type in the address of the webpage you wish to go to.

While you are typing in the webpage address, a live Google search is being conducted in the background where your desired website may come up. You can single tap on that address if it does come up.

When you are done typing in the desired website, then single tap the "Go" button and you will be taken to that webpage. The picture below shows what your screen would look like when typing in the webpage address in the next picture

Just like the browser on your computer, you can set various options such as creating bookmarks among other things. To show the browser options, while having the Android browser open, press the menu key on your phone which looks like this . The option menu will come up like the next picture.

From here, you can do things such as bookmarking the current webpage. To bookmark the current page, single tap the "More" option. Your screen will look like the next picture:

Single tap the "Add Bookmark" option and the "Add Bookmark" prompt will come up as shown in the next picture:

To add the bookmark to your phone, single tap the "OK" button. Your bookmark is now added to your phone. To pull up the bookmarks, single tap the menu key on your phone which looks like this 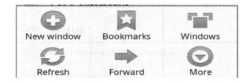. The option menu will come up like the next picture.

Single tap the "Bookmarks" option and the bookmark list will come up as shown in the next picture:

To go to a specific bookmark, single tap the specific bookmark and the browser will then go to that website.

How To Setup And Use E-Mail

Electronic Mail or E-Mail has changed how people communicate from the business world; to families that live far away so that they could easily stay in touch. E-Mail allows you not only to send letters to people; also pictures and more recently, small video clips. Many people already know how to send an e-mail, but not fully utilizing the power of e-mail.

Note – When you purchased your Android phone from your cell phone carrier, you would have to sign up for a Google account, which gives you a Gmail account and the application that you can use for Email if you don't already have an E-Mail account. In that case, skip to page 98. But if you want to use the same email account of your Internet Service Provider for your home computers, you can use that account also with your Android phone.

Before we get into working setting up your email account on your Android phone, we have to Configure Android Mail to work with the E-Mail address your Internet Service Provider has given you to use. The 4 things we will need to get your account to work are listed on the next page:

80

1. Your Internet Service POP Server address (POP stands for Post Office Protocol). It handles E-Mail coming in from people that has sent you E-Mail.

2. Your Internet Service SMTP Server address (SMTP stands for Simple Mail Transfer Protocol). This is used to send messages and forwards that you have written to people.

3. The username you gave the Internet service provider to setup your Internet account with them.

4. Also the password that you gave the Internet service provider to setup your Internet account with them.

If you don't remember or lost the documentation that they gave to you, you are going to have to call the customer service of your Internet Service Provider to get this information, because you are going to need it to get Android Mail® to send and receive E-Mails at all. When you have this information, we can move on to the next step: setting up the e-mail program itself.

To setup the e-mail account for your Internet Service Provider in Android 2.1, Single tap the launch icon from the main screen (You would get to the main screen by pressing the home key on your phone. It look like this) and swipe up or down to find the Android Mail application. Single tap the Android Mail® app, which icon looks like the next picture

Press the menu key on your phone which looks like this

. Your screen should look like the next picture:

Single tap the "Add Account" button and the setup wizard will start to input the information from them as shown in the next picture.

This screen is the first step in the Android Mail® wizard where they ask you to type in e-mail address and your password in the text boxes. Single tap the Email Address box, And the touch keyboard comes up. Type in the e-mail address your Internet Service Provider had given to you. It is usually <u>username@internetprovidername.com</u> .org or .net.

Then single tap the "Done" button on the touch keyboard to close the touch keyboard. Next, single tap the password text box and the touch keyboard will come back up. Type your password in the password box, and single tap the "Done" key on the touch keyboard to close the touch keyboard. Then single tap the "Next" button.

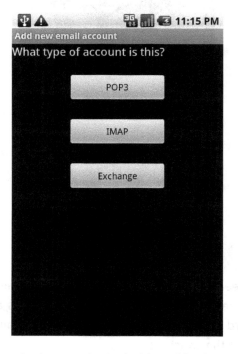

This screen is the step in Android Mail® where they ask you to specify the type of email that you are using. In most cases, it would be POP mail, so single tap the "POP" button.

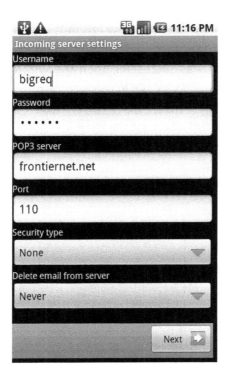

In this window, we are going to enter the POP server information from your Internet Service Provider so the program could connect to the right servers to deliver and to send E-Mail. Single tap the POP3 Server text field and the touch keyboard will come up.

Type your server address in the POP3 server text field, then single tap the password text field and type in your password and single tap the "Done" key on the touch keyboard to close the touch keyboard. Single tap the next button at the bottom of the window.

Your Android phone will now attempt to connect to the email service using the information you typed in. And if the attempt was successful, then the wizard will go to the next screen as shown in the next picture:

In this window, type in the SMTP information from the Internet Service Provider so the program could connect to the SMTP server. Type the SMTP server information in the SMTP Server text box. Single tap the "Require sign-in" checkbox and single tap the next button at the bottom of the window.

This window is where you can tell Android Mail how often to check for new messages. You can single tap the down arrow to change the time. But for now, single tap the "Next" button.

At this point, your e-mail account is just about set up. Now you will have to type in a name for this particular email account. The name could be anything such as "Home or Work Mail" or anything that you could easily identify this account. Single tap the text box and type in the name of the account and single tap the "Done" button. Your Email account is now setup for use.

In Android 2.2, to start configuring Android Mail, single tap the launch icon (You would get to the main screen by pressing the home key on your phone. It look like this) swipe up or down to find the "My Accounts" icon, this icon looks like the next picture.

Single tap on the icon, and the screen should look like the next picture.

Single tap the "Add Account" button at the bottom of this window and the setup wizard will show a window asking you what kind of account that you want to set up. This window looks like the next picture.

In this case, we are setting up an Email account, so single tap the Email option to start to setup your Email as shown in the next picture.

This screen is the first step in the Android Mail® wizard where they ask you to type in your e-mail address and password in the text boxes. Single tap the Email Address box, and the touch keyboard comes up. Type in the e-mail address your Internet Service Provider had given to you. It is usually <u>username@internetprovidername</u> <u>.com</u> .org or .net.

Single tap the password box, type in your password in the password box. Single tap the "Automatically configure account" checkbox to uncheck this option. Single tap the "Done" button on the touch keyboard to close the touch keyboard. Single tap the "Next" button.

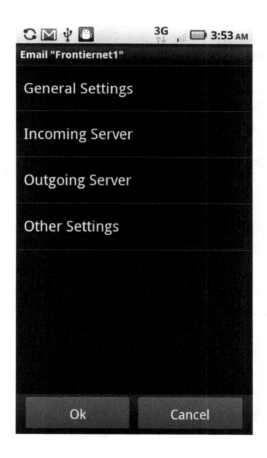

This screen is the step in Android Mail® where you enter the information from your Internet service provider to configure your phone to use their Email account. Single tap the "General Settings" option. Your screen should look like the next picture:

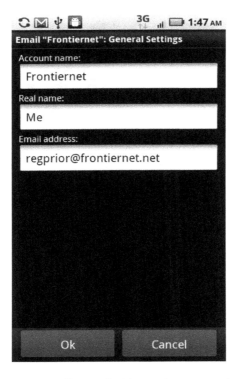

In the "Real Name" text field, single tap on it and the touch keypad will come up. Type your first and last name in this box and single tap the "Next" key. Then single tap the "Done" key to close the touch keyboard.

Single tap the "OK" button to go back to the main information screen as shown in the next picture.

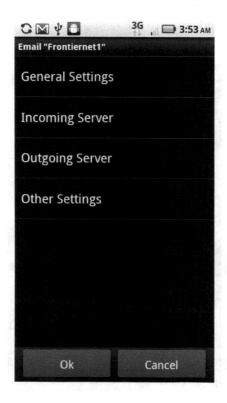

Next, we are going to enter the Incoming Server settings for this email account, so single tap the "Incoming Server" menu choice, and your screen should look like the next picture:

In this window, we are going to enter the POP server information from your Internet Service Provider so that Android Mail could connect to the right servers to deliver and to send E-Mail. Single tap the POP3 Server textbox and the touch keyboard comes up. Type in the POP server information.

Single tap the next button and enter your password in the password box and single tap the "Done" key to close the touch keyboard. Then single tap the "OK" button at the bottom of the window.

Your Android will phone go back to the main email account settings menu as shown in the next picture.

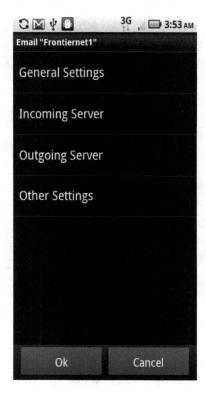

Next, we are going to enter the Outgoing Server settings for this email account, so single tap the "Outgoing Server" menu choice, and your screen should look like the next picture:

In this window, we are going to type in the SMTP information from your Internet Service Provider so Android Mail can connect to the SMTP servers. Single tap the SMTP Server text box and enter the information through the touch keyboard and single tap the "Next" button until the touch keyboard closes. Single tap the "OK" button at the bottom of the window.

At this point, your e-mail account is now set up. On the main settings screen single tap the "OK" button at the bottom right of this window. The phone will start downloading messages from the server within five minutes. You can single tap the "Home" button to go back to the main screen.

To send email using your phone, you would have to go back into Android Mail by single tapping the launch icon from the main screen (You would get to the main screen by pressing the home key on your phone. It look like this) and swipe up or down to find the Android Mail application. Single tap the icon by single tapping on the icon that looks like the next picture:

Email

And the Android Mail app will load and check for new Emails and will look like the next picture:

98

To start composing a new email message, press the menu key on your phone which looks like this  and the email option menu will come up looking like the next picture:

Single tap the "Compose" option to bring up the new message window which looks like the next picture

To create a message, have the person's e-mail handy. Make sure the spelling of the e-mail is correct because your message will not be sent at all if the spelling is incorrect. The steps below are how you would compose the e-mail in Android mail

1. First, single tap the "To:" textbox. The touch screen keyboard will come up. Type the e-mail address of the person that you want to send this message in this text box. Again, make sure that the e-mail address is correctly spelled, or the message will not be sent at all.

2. Secondly, single tap the "Subject:" textbox. Type a brief summary of what this e-mail is about in the textbox.

3. Thirdly, single tap the compose mail textbox on the bottom of the window. Type the actual message in this box.

4. When you are done typing your actual message to this recipient of this message, single tap the send button. Your E-Mail message will be sent to the e-mail address you inputted. That is how you put together and send an e-mail message in Android mail.

Deleting Messages In Android Mail

To delete a message in Android Mail, Single tap the launch icon from the main screen (You would get to the main screen by pressing the home key on your phone. It look like this) and swipe up or down to find the Android Mail application. Single tap the Android mail icon that looks like the next picture:

Email

On the main screen, swipe up or down to find the message that you want to delete, and single tap on it. When the message opens up, it would look like the next picture:

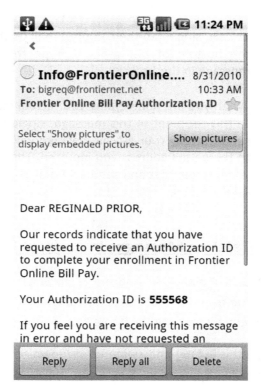

To delete this message, simply single tap on the "Delete" button and the message will be deleted from your inbox. In Android 2.2, the delete button is an icon that looks like a trash can.

Using The Camera To Take Pictures And Video

Another thing that you can do with your Android phone is to use it as a camera or a camcorder to take still photo pictures or shoot short videos. Depending on the size of your memory card in your phone, you can take lots of pictures or shoot hours of video.

To start taking pictures, Single tap the launch icon from the main screen (You would get to the main screen by pressing the home key on your phone. It look like this) swipe up or down to find the Camera application. Single tap the Camera app icon that looks like the picture shown below:

The camera program will come up and looks like the picture shown on the next page:

Android 2.1 Camera Screen

Android 2.2 Camera Screen

To take a picture, simply just point your phone at whatever it is you want to take the picture of and single tap the shutter icon which looks like this

Android 2.1 Shutter Icon

Android 2.2 Shutter Icon

And the camera will focus itself and then take the picture and save the picture to your phones memory card. Another thing you can do with your Android phone is to take videos like a camcorder. To take a short video with your phone, In Android 2.1, you will have to swipe down on this slider to turn your phone to "Camcorder Mode"

Then the phone will turn to camcorder mode and look like the picture shown below:

In Android 2.2, you would have to single tap in the middle of the screen to pull up the menu screen as shown in the next picture.

Then single tap the "Switch To" option to switch the phone into camcorder mode.

To take video, single tap the red record button. It looks like this. ⊙ The phone is now recording video and would look similar to the next picture:

To stop taking video, single tap the stop button. That button look like this ⬜. Your phone will stop recording video and save the video on your memory card.

How To Download Pictures To Your Phone

You may have pictures saved on their computer that you may want to show to other people or view for yourself from time to time. You could get one of those keychain digital photo frames that look like the next picture:

107

But the truth of the matter is that the quality of the pictures downloaded to the frame leaves a lot to be desired and the software on some of them that are used to download pictures from your computer to the frame are not very user friendly.

The more logical solution is to use your phone as a portable picture frame. And the way to download pictures off of your computer to your Android phone cannot be any simpler than the instructions we are about to get into now.

The first thing you will have to do is to plug one end of the USB cable that came with your phone into your phone and the other end into your computer. Then put your finger on the notification bar (The same toolbar where the clock is located), single tap and hold your finger on the screen until the notification bar looks like the next picture:

Then swipe down to show the notification menu as shown in the picture below:

Your phone should show in this menu the option "USB Connected. Select to copy files to/from your computer". Single tap this option. In Android 2.1, The phone will show the USB connected prompt asking you to single tap the "Mount" button to connect your phone to the computer as shown in the next picture.

Single tap the "Mount" button and your phone will connect to your computer. In Android 2.2, the USB connection prompt will come up like the next picture:

Single tap the "USB Mass Storage" option, and then single tap the OK button and your phone will connect to your computer. Your computer will then show your phones memory card as a removable hard drive on your computer as shown in the next picture:

To transfer your pictures from your computer to your Android phone, I suggest that if you don't already have Google's Picasa installed on your computer, that you download and install it from http://www.picasa.com. After you have downloaded and installed Picasa onto your computer, the installer will leave an icon on your desktop you would double left click on to open the program. The icon looks like this:

After you double left click on the icon shown above, Picasa will then open up to the main screen that will look like the next picture:

Picasa will then search your computer for pictures and load and organize them on this main screen. As you can see, Picasa is divided into several parts. We will go over each part in detail in the next section.

113

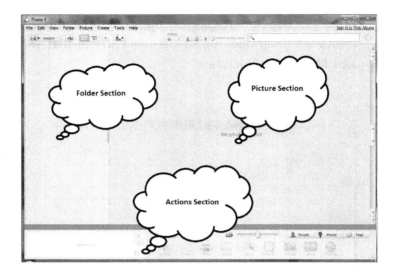

Picture Section - This is the section where previews of pictures will show up when you single left click on a folder in the folder section.

Folder Section – This is the section where you will organize your pictures into albums for easy and quick access.

Actions Section – This is the section where you would tell Picasa what you want to do with a picture or album you have selected.

114

To transfer pictures to your Android phone, simply select one of your pictures from an album by single left clicking the picture in the picture section, and holding the left mouse button down. Drag the picture down to the selection tray in the action section of the main window and let go of the left button on your mouse. It looks like the next picture:

When you let go of your mouse button, your picture should then appear in this window with a green circle on the photo looking like the picture below:

This will let you know that the picture was successfully inserted into the selection tray. Then go through the rest of your albums and select other pictures you wish to transfer and put them into the selection tray the same way. When you are done selecting pictures and want to transfer them to your phone, single left click the export button on the actions section. The export menu will come up looking like the next picture:

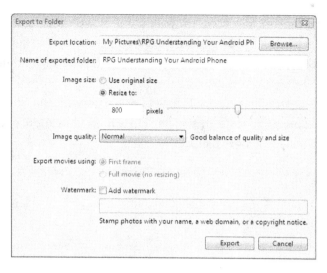

When this menu choice comes up, you are going to change the Export location field to be the drive letter of your Android phone's memory card. To do that, single left click the "Browse" button and the menu prompt will come up like the next picture:

Single left click the arrow right next to the computer label and all of the drives on your computer will show up like the picture shown above. In this case, the Android phone's memory card drive letter is the (F:/) drive, so we would single left click the (F:/) drive once and then single left click the "OK" button to select it. Your computer screen should look similar to the next picture:

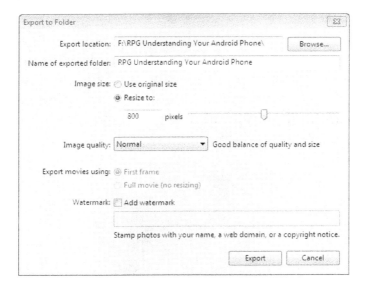

At this point, you could single left click the "Name of exported folder" text field and give this album a different name, or you could leave it as is. Now you would transfer the pictures to your phone by single left clicking the "Export" button. Your pictures will now transfer to your phone.

When you are done copying all of the pictures that you want to your phone, disconnect the USB cable from your computer and you should be able to single tap the gallery app in the launch menu (you would go there by pressing the home button on your phone) which icon looks like the next picture:

The transferred pictures should show up where you can single tap on them to show up full screen on your phone as shown in the next picture. I had uploaded a folder called "Benjamin6Mos" to my phone, as you can see it below in the gallery.

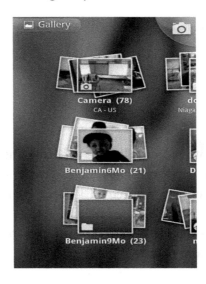

Single tap on that album, and all the pictures in that album would show up on the screen. You can show your downloaded pictures in a slideshow by single tapping a single picture in the album. The picture will show up onscreen with the option on the bottom left part of the screen to single tap on the slideshow button, where all your pictures in that album will show in a slideshow.

How To Download Music To Your Phone

To download music to your phone, the first thing you will do is to plug one end of the USB cable that came with your phone into your phone and the other end into your computer. Then put your finger on the top toolbar (The same toolbar where the clock is located), and single tap and hold your finger on the screen until the top toolbar looks like the picture below:

Then swipe down to show the notification menu as shown in the next picture:

119

Your phone should show in this menu the option "USB Connected. Select to copy files to/from your computer". Single tap this option. The phone will show the USB connected prompt asking you to single tap the "Mount" button to connect your phone to the computer as shown in the picture below.

Single tap the "Mount" button and your phone will connect to your computer. In Android 2.2, the USB connection prompt will come up like the next picture:

121

Single tap the "USB Mass Storage" option, and then the OK button and your phone will connect to your computer. Then your computer will show your phones memory card as a removable hard drive on your computer as shown in the next picture:

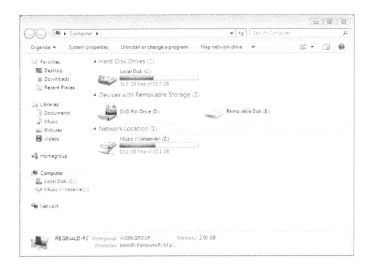

After your computer recognizes the phone, you can use Windows Media Player to transfer songs to your phone. One of the features of Windows Media Player is the option of being able to transfer your music from your computer to your Android Phone.

But before we can transfer music to your Android phone, first we have to download music to your computer to transfer to your Android phone. To get started working with Windows Media Player, first we are going to download music from your music CD's to your computer. And to do that, get your favorite music CD's together to import into your computer.

Open up Windows Media Player to the main screen and insert one of your music CD's into your computer.

Windows Media Player will read the CD and the songs from the CD will appear in the song list section of the program as shown in the next picture:

The next step to transferring a music CD to your computer is first making sure that Windows Media Player transfers the music to your computer into the MP3 format. This format is most compatible with your phone.

And to do that, depends on which version of Windows Media Player you have installed on your computer. In Windows Media Player 11, which is installed for Windows XP and Vista, move your mouse cursor to the down arrow under the "Rip" button and single left click on it. A submenu will show up as shown in the picture below.

Windows Media Player 11 Screen

125

Under the format submenu, make sure you have the MP3 format check marked as shown in the picture on the last page. With Windows Media Player Version 12, which is only for Windows 7 at the time of the writing of this book, single left click the "Rip Settings" button.

Under the format submenu, make sure you have the MP3 format check marked as shown in the picture below.

Windows Media Player 12 Screen

After that, single left click the CD icon on the library section of Windows Media Player. The songs on the CD will appear in the songs list section with a check mark next to each one of them. If you want all of the songs on this CD to be transferred to your computer, leave the check marks there. If not, uncheck the songs you don't want to transfer by single left clicking the check mark of each song to unselect them from the transfer list.

When you're ready to download the music CD, single left click the "Rip" button. Then your CD will download to your computer. Repeat the last instructions for each additional CD you want to import into your computer to transfer to your phone.

The next step in the process to play your music on your Android phone is to actually transfer music from your computer to your phone. And to do that, single left click the arrow next to the music menu in the Library section on the main menu. A submenu will come up with three choices, Artist, Album & Genre. Single left click on Artist.

All of the music that you have downloaded onto your computer will come up in the song list section, arranged by Artist. Double left click an artist to show songs and individual albums you have from that artist on your computer. To transfer an individual song to your phone, move your mouse cursor to the song and single left click and hold the left mouse button down.

Drag the song to the sync list where it says "Drag Items Here" and let go of the left mouse button. Repeat for each song you want to transfer. Then move your mouse cursor to the "Start Sync" button and single left click on it to transfer the music to your phone.

128

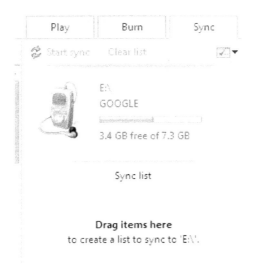

Play	Burn	Sync

Start sync   Clear list

E:\
GOOGLE

3.4 GB free of 7.3 GB

Sync list

Drag items here
to create a list to sync to 'E:\'.

To transfer a complete album, move your mouse cursor to the album cover and single left click and hold the left mouse button down. Drag the album cover to the sync list and let go of the left mouse button. Repeat for each album you want to transfer. Then move your mouse cursor to the "Start Sync" button and single left click on it to transfer to your Android phone.

When you are done copying all of the music that you want to your phone, disconnect the USB cable from your computer and then the music transferred to your phone will be ready for playing.

129

Using The Gallery To Delete Pictures And Video

There are times when you want to remove a picture or video clip from your phone to free up space on your phone. To delete a picture or video clip from your phone, open the Gallery app from the launch menu from the main screen (You would get to the main screen by pressing the home key on your phone. It look like this) swipe up or down to find the gallery app as shown in the next picture.

Single tap the icon and your pictures will show up where you can left or right swipe in the gallery and single tap a album to show pictures as shown in the next picture.

To delete a picture or video clip on your phone, first open up the album where the picture you wish to delete are located by swiping left or right through that gallery and single tapping on the album to show all of the pictures in the album as shown in the next picture:

To select individual pictures to delete, press the menu button on your phone which look like this [icon] twice and all of the pictures in the gallery will show at the top right of each one of them greyed out checkboxes where you would single tap on each individual picture you want to delete.

When you have selected all of the pictures you want to delete from your phone, single tap the "Delete" option at the bottom of the screen. When you single tap the delete key, a prompt will come up wanting you to tap the "Confirm Delete" option to finish deleting the picture from your phone as shown in the next picture:

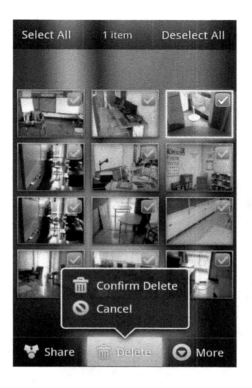

In Android 2.2, to select pictures that you wish to remove from your phone, press the menu button on your phone, and the menu comes up where you would single tap the "Select Items" menu option as shown in the next picture:

After you single tapped the "Select Items" option, check boxes will display on each photo where you can single tap individual pictures to select them as shown in the next picture

After single tapping the desired pictures or videos to remove from your phone, press the menu key on your phone and the menu will come up as shown in the next picture:

Single tap the "Delete" option, and the delete prompt will come up as shown in the next picture:

Single tap the "OK" button and the pictures will be removed from your phone.

Using The Music App To View And Play Your Music

In a previous section, "How to Download Music to Your Phone" we went over how to transfer music from your personal CD's to your computer and from there, on your phone. Now you want to play the downloaded on your phone.

You would do that using the music app. From the launch menu (You would get to the main screen by pressing the home key on your phone. It look like this), swipe up or down to find the music app icon, which looks like the next picture:

Single tap on the icon, and the music that you have downloaded to your phone should show up where you can single tap on them to play on your phone as shown in the next picture.

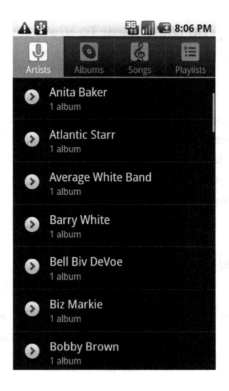

To see a song list from a particular artist, single tap the artist's name. In this case, I will tap the artist Barry White. All of the albums that I have downloaded from Barry White will come up as shown in the next picture:

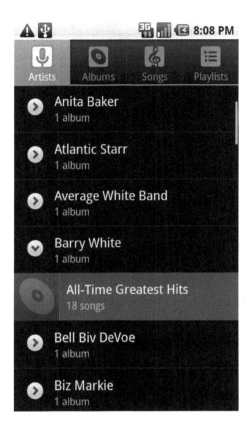

Then to see the songs from a particular album, single tap the album and the song list will show as shown in the next picture:

When the song list comes up, you can swipe up or down to view all of the songs and then single tap a particular song to play it. When you do that, your screen will look like the next picture:

On this screen, you can rewind, pause and go to the next song by tapping on the appropriate buttons. The buttons are shown below:

 This button is the pause/play button. When the song is currently playing, this button will turn to the pause symbol. When a song is paused, then this button will turn to the play symbol.

 This button is to cycle to the next song on The current playlist

This button is to go back to the previous song on the playlist.

To go back to the main music screen, press the back button on your phone which looks like this and then you would go back to the song list.

Note – If you want to delete a song from your phone, while the song is playing, press the menu key on your phone, and the song menu will come up. Single tap the "Delete" menu, and the song will be deleted from your phone.

Chapter Three:

Using The Android Market To Add Functionality To Your Android Phone

Getting To The Android Market

A lot of buzz words have been circulating about smartphones for a year or so. But the word "Apps" has been a big one for sure. Apps are short for Application and share the same meaning as with regular computers.

Applications are programs designed to perform a function or suite of related functions of benefit to an end user such as email or word processing. On an Android phone, you would install your applications from the Android Market.

To get to the Android Market, on the main screen (You would get to the main screen by pressing the "Home" key on your phone. It would look like this [image]), you should see the Market icon. Single tap the Market Icon which looks like the next picture:

And the Android market main screen will come up looking like the next picture:

Downloading And Installing Apps

To install an app on your phone from the Android Market, there are two ways to do it. The first method is that if you know the name of the app, single tap the magnifying glass in the upper right part of the screen and a search box will come up like the next picture.

Type the name of the app in the search box and single tap the hourglass icon to search the market for the app. If the app is available on the market, then it will show up in the results box where you would single tap on it to install on your phone.

The second way to install apps on your phone is to actually look around on the Android market for apps. Keep in mind that there are hundreds of thousands of apps on the Android market and growing every day.

The easiest way to do that is when you are at the Android Market main screen, single tap the "Apps" button. Then the apps menu screen will come up arranged by categories that you would select for different type of apps as shown in the next picture:

145

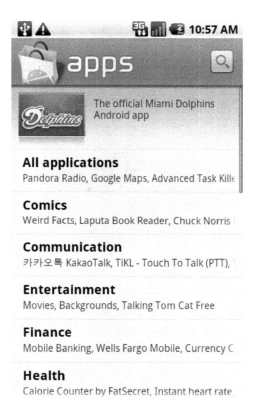

So single tap the category that interest you such as multimedia, and all of the multimedia apps in the Android market will come up as shown in the next picture:

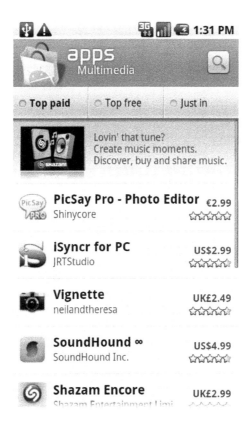

You would swipe up or down to take a look at all of the apps in this category, and when you come across one that interests you, single tap on it to view a description of the app and the button to install it as shown in the next picture:

See PicSay Pro for much more features and stickers.

Powerful and award winning photo editor. Color-correct your pictures and add word balloons, titles, graphics, and effects like distortion. All in a fun, intuitive, and easy-to-use interface.

HTC Incredible users need an SD Card!

Version 1.3.0.8   0.89MB

Note – Some apps are free and some are ones that you have to pay for. And the way that you can tell if an app is free or pay is that in on the same line of the app, it would either list "Free" or the cost of the app in US dollars as shown in the next picture:

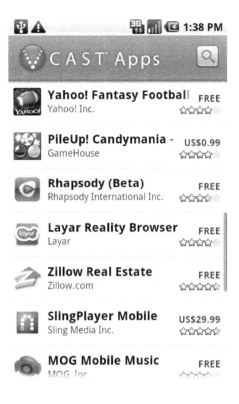

And if you install the apps that are not free, usually it would go on your next cellular phone bill as an additional charge.

Another note about apps – Periodically the companies or persons who created these apps that are installed on your phone periodically releases updates to these apps. To update them, an update icon will appear on your notification bar with the Android market logo.

You would swipe down the notification bar, and a menu choice would state that they are updates available for your phone. Single tap that menu choice and all of the updates for the apps will come up as shown in the next picture:

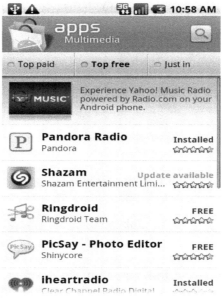

The apps installed on your phone that has an update available will be flagged as "Update Available". To update that app, single tap on the app and a menu will come up for that particular app. Single tap the update button and another prompt will come up stating that this program will be updated and your existing files will be saved. Single tap the OK button and the app will be updated.

Uninstalling Apps That You Don't Use

From time to time, you may have installed a lot of apps and need to clear some space on your phone for one reason or another. The best way to clear space on your phone is to remove the apps that you don't use that often.

To remove the apps, go to the main screen (You would get to the main screen by pressing the "Home" key on your phone. It would look like this ⌂), press the menu key, which looks like this ☰ and the options will come up like the next picture:

Single tap the "Settings" option and the settings menu will come up like the next picture:

Single tap the "Applications" menu choice and the applications menu will come up as shown in the next picture:

Single tap the "Manage Applications" menu choice and all of the apps that are installed on your phone will come up like the next picture:

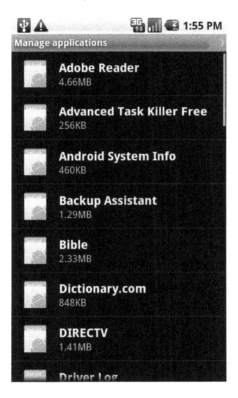

To uninstall an app, swipe up or down to find the app that you want to remove and single tap on it to bring up the uninstall menu as shown in the next picture:

On this menu, single tap the uninstall button and then the app will be removed from your phone.

Chapter Four:

Smartphone Security

As convenient and easy it is to do many things on your smartphone, the fact remains that smartphones carry a lot of personal information on them where it could be dangerous if your phone ever gets stolen or someone else who is using your phone does something on an application where it could adversely affect you in one way or another.

Here are some things you can do to lessen your risk of these things to occur.

1. Go to the Android market and install the lookout security app. Then either go to https://www.mylookout.com/ and sign up for an account, or do it through your Android phone.

   This application includes an antivirus and malware program, a backup utility to make sure that you don't lose anything on your phone and also if your phone is lost or stolen, you can locate your phone or wipe your phone so that no one would be able to have access to your information. Also you can back up your contacts and other data from your phone.

2. The second thing you can do to protect your data is to set an unlock pattern for your phone. You would do that by going into the settings menu by going to the main screen (You would get to the main screen by pressing the "Home" key on your phone. It would look like this), you would press the menu key, which looks like this and the options will come up like the next picture:

Single tap the "Settings" option and the settings menu will come up like the next picture:

Single tap the "location & security" menu option and the screen should look like the next picture:

To set an unlock pattern, single tap the "Set unlock pattern" option (In Android 2.2, it would be labeled "Change Screen Lock"). Your screen will come up like the next picture:

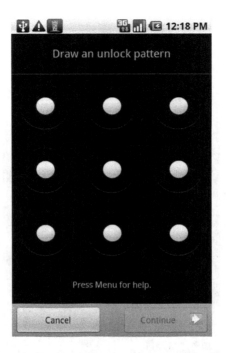

Use your fingers and draw an unlock pattern. Remember to create a pattern you can easily remember because if you forget the pattern, you cannot unlock your phone and would have to go to your carrier to reset your phone which case, would wipe everything out on your phone!!!!

When you have created a pattern, single tap the "continue" button. You will be prompted to draw the pattern again. Draw the pattern again and single tap the "confirm" button to set the pattern on your phone.

Now your unlock screen will look like the next picture:

If you don't want to keep the unlock pattern and want to swipe in, go back into the settings menu by going to the main screen (You would get to the main screen by pressing the "Home" key on your phone. It would look like this ⌂), you would press the menu key, which looks like this ▤ and the options will come up like the next picture:

Single tap the "Settings" option and the settings menu will come up like the next picture:

Single tap the "location & security" menu option and the screen should look like the next picture:

To unset a unlock pattern, single tap the "Require pattern" option. Then you would be asked to draw the unlock pattern to disable the unlock pattern.

3. The third thing you can do for data's safety is make sure that you know where you put your phone at ALL TIMES!!!

Using a combination or all of these methods listed above will keep your data and your smartphone safe from cyber thieves and other threats.